How to Earn $100.00 an Hour, Without a High School Diploma or a GED as a Non-Licensed Plumbing Drain Cleaning Technician

Doing Basic Home Drain Cleaning Service, Installation and Repairs

BROAD Vision Publishing

BROAD Vision Publishing
"Narrowing the gap between <u>you</u> and <u>success</u>!"

How to Earn $100.00 an Hour, Without a High School Diploma or a GED as a
Non-Licensed Plumbing Drain Cleaning Technician

Doing Basic Home Drain Cleaning Service, Installation and Repairs

This book is designed to provide accurate information on the subject matter covered. However, it is sold with the understanding that the publisher is not engaged in rendering any employment guarantee, legal advice, accounting practice, or professional services. If any legal counseling or any other professional assistance is required the benefit of competent professional individuals should be sought to assist with any specific questions or concerns.

How to Earn $100.00 an Hour, Without a High School Diploma or a GED
As a Non-Licensed Plumbing Drain Cleaning Technician

Plumbing can be one of the most challenging and satisfying professions of all the building trades. Variation of design, layout, and installation can become a stimulating challenge for the professional plumber as well as handy men and women.

In my practical guide to becoming a plumbing drain cleaning technician, fixture installation and service repair business, you will discover how profitable the residential repair market can be. Often-basic home maintenance can be overwhelming for a homeowner, which can generate hundreds of dollars being spent on basic home repairs and installation everyday.

Basic repairs that can really add up are clogged drains, upgrades, leaking kitchen and bathroom fixtures. Have you ever noticed how happy your plumber is when greeting you? At an average cost per hour of $125.00, plus material cost, homeowners and businesses can spend a lot of money really fast! The more you are able to assist the customer the greater the benefit will be to your growing business. Plumbing drain cleaning, fixture installation and repair can be simple with the correct plan and preparation. You can begin to build large profits in a home repair service venture. However, some repairs and installations may require calling in a professional plumber.

Lack of knowledge about the uniform plumbing code has left many would-be plumbing drain cleaning technicians and handymen/women disgusted, frustrated, and confused.

Fortunately, this is not a study guide in mastering plumbing theory or code. Nor will this guide enable you to complete a weekend bathroom remodel. However, the Uniform Plumbing Code is an excellent reference guide to study for a career as a licensed journeyman plumber and plumbing contractor.

The Uniform Plumbing Code (UPC) is a regulated guide designed to provide direction to the professional plumber. It explains important design and installation principals and practices established for plumbing and pipe trades. In the chapters ahead it should be less confusing for the beginning trades men and women to read this guide. Rather than reading and comprehending the Uniform Plumbing Code that is hard to understand. This guide is designed to assist anyone who is interested in learning the fundamentals of the plumbing and drain cleaning service business.

For the would-be professional plumber, plumbing theory will require a general understanding of the uniform plumbing code. You will soon discover parallel similarities in safety and sanitation practices that most jurisdictions enforce inside the uniform plumbing code. The National, State and local authorities regulate the Uniform Plumbing Code.

BROAD Vision Publishing will soon be releasing an instruction manual on How to become a "Certified Journeyman Plumber," and preparing for your "C-36 Plumbing Contractor Exam" Complete with example test questions and answers.

History of Plumbing

Plumbing systems were developed to prevent a mass pandemic of diseases and plagues. Early social acculturation soon recognized the importance of proper sanitation. Massive pestilence in communities caused devastated diseases carried through waterways and insanitary disposal methods. Archaeologist revealed aqueduct systems that date back as early as 4300B.C. Palaces outfitted with elaborate water and drainage systems have been discovered. Ancient clay plumbing fixtures have also been discovered. England sanctioned the pristine trade guild in the 12th century, under the sovereignty of Queen Elizabeth. England also culminated the first Master of Plumbers association and incorporated it into the College of Heralds of London. The British were the first to introduce lead to the Romans. The Romans used lead that constructed water channels, vats, and disposal fields. They referred to the pliable material as 'Plumbum'.

The Latin word for lead is "plumber" or "plumbing." The modern plumber was referenced as the "plumbaris" which means worker of lead. America is symbolic of conventional plumbing and sanitation techniques. Developing proper sanitation designs was a very tedious process. New York installed the first underground sewer in 1782. Chicago is credited with having the first functional sewage system constructed in 1855.

Modern conveniences like a restroom, potable water, drinking fountains, and functional sewer systems can be taken for granted. Lets' not forget outhouses, wooden wash tubs, and open water wells were some of the luxuries we enjoyed not too many years ago.

Financial Freedom is Now Within Your Reach! What is Your Time Worth?

Welcome again,
You have just taken the first steps on a journey toward unlimited financial freedom. Self-employment is the best investment you can make. In today's economic challenge small businesses will be instrumental in creating new jobs throughout the country. By becoming an entrepreneur you can use specialized knowledge and skills that can assist in building a lucrative and profitable business. You can start earning $100.00 an hour!

The key essentials to any successful business endeavors are preparation and effort. It takes focused decision-making, that will guide you in building a strong profit margin and employment independence!

- "Skilled Plumbing Drain Cleaning Technicians and, Journeyman Plumbers, will undoubtedly remain an exceedingly HIGH EMPLOYMENT DEMAND. It is a PROMINATE and REWARDING PROFESSIONAL TRADE."

(BROAD Vision Publication 1992)

There are over 245 million homes, apartments and commercial structures throughout the nation. Any building connected to a sanitary disposal system must be installed and maintained in accordance with provisions of the uniform plumbing code (UPC). The sections number 304.0 of the International Association of Plumbing and Mechanical Officials (IAPMO) Uniform Plumbing Code states: **Any building used as occupancy shall be connected to a sanitary plumbing System**

"All plumbing fixtures, drains, apparatuses and appliances used to receive or discharge liquid wastes or sewage, shall be connected properly to the drainage system of the building or premises, in accordance with the requirements of this code."
(Uniform Plumbing Code TM, 1997 Edition)

Drain cleaning technicians and licensed journeyman and woman plumbers seldom experience shortage of work in their professional trades. A drain cleaning business can earn an average income between $50,000-$150,000 annually doing drain and waste maintenance.

Licensed Journeyman and Woman plumbers seeking professional employment as plumbers can be offered lucrative salary and benefit compensation. Some can earn an average annual income of $65,000-$90,000 annually. A C-36 licensed plumbing contractor can experience abundant income opportunities.

In a professional industry that is largely occupied by men, major plumbing and mechanical companies throughout the country are aggressively seeking qualified woman drain cleaning technicians, and journeyman women plumbers. Women are discovering new and interesting challenges as plumbers as well as handsome salary rewards.

Women are strongly encouraged to pursue a career in plumbing drain cleaning and plumbing/pipe trades. Woman business owners have established a competitive and profitable financial edifice with creditable clientele networks by simply meeting the market demand.

Working smart and efficient by preparing the groundwork required in the beginning phase of building your own financial security. I will show how you can become a qualified candidate following my three simple rules:

Rule number 1: Your time is money. Every service call can become a cash opportunity for you. Always remember, when a customer is pleased with your professional service, you will benefit your business and expert reputation tremendously. This will prolong profitable consumer relationships.

Rule number 2: Always provide respectful and courteous service. An indirect contributor to your business success will be through customer referrals as well. Referrals from family, friends, neighbors, and business associates can become just a few key components that will return huge dividends to your business.

Rule number 3: Establish a list of wholesale suppliers and home hardware retailers. Also, set up a business credit account. Even with no prior credit history, or an unfavorable credit chronicle most retailers will work with you on a merit covenant. Honor the trust that is granted onto you when you're entered into that type of purchase agreement. Ultimately, it is a tremendous advantage in having an ability to purchase parts on credit when you need to procure material and equipment. By following those three simple rules, you should be ready to begin building your own financial future.

- Before you continue reading this book, answer these several questions I often ask:

Have you ever wanted your own business?

Are you ready to earn some serious cash finally?

Do you want to be your own Boss?

If you answered "yes" to any of the questions I have asked, you are definitely ready to embark upon a financial future that can be filled with unlimited opportunities. You can plan your own work schedule. Work the hours you choose. Also, pay yourself handsomely for what your time is really worth now!

This book is also designed to assist the layman as well as individuals with some acquaintance in basic residential drainage system design and maintenance repair work. It can also be a great reference manual for individuals with some moderate plumbing trade familiarity.

The technical information inside of this guide is examples to assist you and simplify how to perform some general home plumbing maintenance and repairs. You will learn how to clear a clogged drain obstruction, install a kitchen faucet, lavatory faucet, and repair a leaking faucet or hose bibb.

These basic skills can sustain a steady customer clientele and, generate some very profitable account revenue for your business. You will also be developing strong trade awareness and dexterity that can build invaluable experience and references if you apply for a journeyman plumber license exam in the future.
You can confidently position yourself to start earning $85.00-$100.00 per hour immediately in your new business venture as a plumbing drain cleaning technician.

When performing professional service in maintenance and repair of building sanitary waste and disposal systems. You need to start familiarizing yourself with trade terminology, tools and materials. You will be selling your knowledge and skill.

I will list several basic repair tips that can be performed efficiently and profitably so you can go out and **$ell Your Time**. Nationally, drain cleaning service companies have made millions of dollars performing clogged drain and sewer repair.

So, let's get started first by developing a plan. It will be necessary for you to review the step by step information I have prepared for you. These steps are essential before you are ready to open for business. We will begin by gaining a thorough understanding of some plumbing system and trade terminology. Also, you will need to become acquainted with the correct use of various hand tools, equipment, and how to properly approach a repair.

Basic maintenance and plumbing drain cleaning repair can start to generate income revenue very fast. Some believe plumbing is nasty and dirty. Often assuming plumbers hate their job so much that they overcharge for everything. Well, it's quite the opposite. "I discovered that most plumbers enjoy their job very much. They take pride in having satisfied customers." (E. L. Broaden, owner Broaden plumbing est.1992). A satisfied customer in most cases will return for more business. Also, they give referrals to your establishment when additional maintenance work or service is required. That can add up substantial profit to your annual portfolio.

To better understand some of the methods used for plumbing repairs we will focus our attention on some simple basic home services that can offer some valuable skill and trade knowledge as well as large profit to your growing professional service. I will show you **How to Earn $100.00 an Hour as a Non-Licensed Plumbing Drain Cleaning Technician.**

A drain cleaning service can potentially gross $150,000 annually. By simply clearing clogged drains. For some plumbing service companies, clearing clogged drains can become more than 75% of their annual service request. That is just one non-licensed plumbing jobs often required in an endless array of plumbing engineering and design. You can earn a considerable amount of money doing just that specialty. Don't you agree?

Whether you decide to do what is referred to in the trade as "specialty work" or not, you can build your entire home repair service around one or even several different repairs and finish work. Starting up what is referred to as a "Handyman" business such as: drain cleaning, fixture installation, faucet repair/installation, hose bibb repair, painting, light carpentry, and many others can be a profitable service. I believe the concept still encompasses the same common goal: "Maximize your earning potential." However, I will focus on what have most benefited my career decision and me personally.

A specialized home repair business can be the most profitable and efficient way to streamline a residential plumbing drain cleaning business. Since "Time is Money" a specialty service does not require a large staff of employees or a lot of costly overhead, so you can maximize your profit margin. A plumbing drain cleaning service charge can average around $85.00 to $100.00 per hour starting.

$PECIALIZE and MAXIMIZE it pays!

I began my career as a plumbing drain-cleaning technician in 1989. November 17, 1987 was the start of a new beginning in my life. After six years working in aerospace fabrication as a Numerical Control (NC) Machinist "company down sizing." Would end my fabrication career. I was being relieved of my duties. I never thought it could or would happen to me. I became an unfortunate casualty of the "Aerospace" budget cuts and was placed on a permanent layoff status.

The reality of not having a job or an income was overwhelming and frightening all at the same time. What do I tell my family, the mortgage company, creditors, etc. After all that they weren't down sizing. To get paid $35,000 a year was a lot of money for a machinist, so I thought. Boy was I fooled.

Days turned into weeks, weeks turned into months. Unemployment benefits were running out and I began to wonder if I would ever work in a professional gratifying career again. Would I ever find another profession that would provide a living for my family and me?

One day, an ex-aerospace colleague stopped by my house. While talking we both shared some of the common frustrations you can experience in job searches. We had both been trying to find work in the rapidly fading aerospace manufacturing industry. My efforts seemed hopeless at the time. Totally by chance he invited me to join him on a few service repair jobs. He was the owner of his own business! A very busy home repair service. To my astonishment after my first day assisting him, doing some basic home handyman repair work. I envisioned that my financial future was destined to become profitable and secure. I was right! "I have not been out of work since that day."

Five weeks after that I started my own business. Now, because I have experienced professional and financial successes in my reformed career choice, I am excited to share this information with you. Since I've started my residential drain-cleaning repair service, I have obtained my journeyman plumber license and a C-36 State Contractor Plumber License. By doing that, it has enabled me to compete for home construction and remodel contracts, commercial business repair agreements, plumbing service work, professional trade and community support with monetary independence.

Starting my own home repair service as well as becoming a licensed journeyman plumber is "one of the most personally rewarding decisions I have ever made."

I will discuss how you can prepare for your Journeyman Plumber exam and C-36 State Contractor Plumber License in an additional publication coming soon. Now, let's get you started on your way to financial freedom and independence.

• Plumbing Terminology

Proper trade terminology is important. Here are some of the definitions used in plumbing:

Accessible: when used in reference with plumbing and related fixtures shall mean having access to but may require the removal of a panel or cover.

Readily accessible: mean having "direct access" without requiring the removal of any panel or cover.

Access panel: a constructed door hinged panel with a locking device. Normally mounted in a wall to conceal plumbing valves and pipe. Sizes usually between 12 inches, 18" and 24" are most commonly used.

Administrative authority: an official State, County or City department agency or law administrator.

Aerator fitting: a device used to combine a mixture of air and water.

Backwater valve: a device used to prevent reverse flow of waste drainage. Installed within a building sewer line.

Ball cock valve: a device found inside a toilet tank used to regulate and maintain proper water fill level inside fixture.

Bathroom: any room equipped with a bathtub or shower.

Branch: any part of a piping system that is not used as the main service pipe.

Branch line: a supply line that connect more than one fixture to a main supply riser or branch pipe.

Building drain: the lowest part of a drainage system used to receive and carry waste from soil and waste pipe through to building sewer.

Building sewer: Horizontal piping of a drainage system that extends from the end of a building drain and receives the discharge form the building drain and carries it out to a public sewer, private sewer, individual sewage disposal system, or other point of disposal.

Caulking: a method of connecting bell and spigot pipe using a lead and oakum seal.

Check valve: a valve designed to allow flow in only one direction.

Clean out: a threaded plug or cover that can be removed to access and clean inside pipeline.

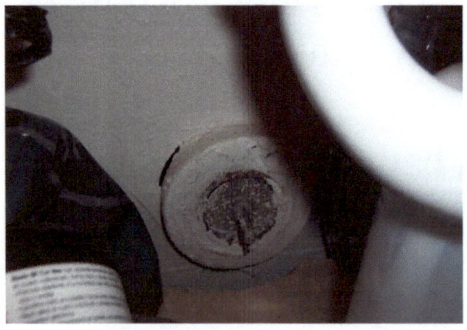 

Developed length: the total length measured along the centerline of the pipe and fittings.

Drain: pipe that carries soil and liquid waste from building drain lateral into building sewer.

Electrolysis: corrosion effect from an electric current usually between two dissimilar metals.

Fixture: a receptacle that is connected to the building drain and a sewage disposal system.

Gallons per minute (GPM): a recorded flow of liquid that passes transitory through a metering device every minute.

Potable water: Water that is satisfactory for drinking, cooking, and domestic purposes and meets the health authority requirements having jurisdiction.

Septic tank: A watertight receptacle that receives the discharge from a building drainage system. Designed and constructed to retain solids and digest organic matter through a period of detention and allowing the liquids to discharge into the soil outside of the tank through a system of open joint pipe or a seepage pit that meets the regulating plumbing code requirements.

- You will become familiar with the plumbing definitions I have listed here and many others during your career as a non-licensed plumbing drain cleaning technician, journeyman and woman plumber.

- Proper terminology and description is a significant benefit when preparing an estimate, material list, and assisting a prospective customer.

• Estimating

Estimating a repair can be done easily by preparing an itemized cost sheet. A price list will provide a quick reference for you when arranging a quote for a customer prior to scheduling repair work. With some experience you will be able to approximate the manual labor and material cost in just minutes. For example: A customer would like to replace an old dripping kitchen faucet.

Installing a single lever kitchen faucet:
One Hour to remove the old faucet $85.00, one hour to install new faucet $85.00, plus kitchen faucet material cost $130.00 equal $300.00 for the installation and material. Start to finish.

You can profit over $170.00 in less than 2 hours. Remember "Time is Money" plan your time to maximize your financial future. You can begin to see your earning potential by use of proper time and money management.

◆ TOOLS & EQUIPMENT

Tools and equipment will be the most important investment you will make starting your home repair service. "You're only as good as the tool." Don't be cheap when outfitting your business tools. Purchase a few necessary hand tools and equipment when you begin your home repair service. However, do not over do it. You can consistently add to your tool inventory over time. You will be able to find excellent purchase opportunities and deals on tools inside trade publications, supply houses, and referral manuals if you look for them.

You might also find the perfect pipe wrench or drain auger at very affordable prices at yard sales, flea markets, and discount centers. New tools are always preferred but are not vital to your business success. Effort will determine success.

Make gadget and equipment purchases that are within your tool allowance. Keep in mind, I recommend "good tools" not the most expensive, or that brand new state of the art do it-all wonder tool. Good hand tools can last you many years if they are properly cared for. Don't shy away from deals on used plumbing equipment and hand tools. You can find some really good deals if you look. Here are several tools you will need to get started in your home repair service.

- **TOOLS**

Safety glasses or goggles (ALWAYS USE SAFETY GLASSES)

When working beneath cabinets and confined spaces, it can become easy for debris to enter into your eyes. "You can't fix what you can't see" please keep your eyes safe. Always wear eye protection.

Leather work gloves

Drain cleaning and other repairs may require you to handle pipe and equipment with sharp, jagged edges or burrs. Protect your hands from cuts and injuries by use of sturdy gloves when necessary.

Drain auger "Ugly Glove"

"Ugly gloves" is a specially designed glove with raised plastic granules on it. Its concept is to prevent hand injuries that might be caused to drain cable operators when using power driven auger equipment.

Measuring tape 25'-0"

Used for measuring pipe and distance expansions.

Comfortable work cap

Comfortable headgear can keep your head cool and clean.

Pipe wrenches 12,"14,"18,"24"

Various sized pipe wrenches can assist with assembly and disassembly of dissimilar size pipe diameters.

Basin wrench/with extension

Basin wrenches are used to access retaining nuts at the base of kitchen and lavatory faucets.

Adjustable pliers

Quick adjustable pliers can be used for assembly, dismantling, and securing materials on various repairs.

Standard Flathead screwdriver

A flat head screwdriver is used to install, and remove slotted wood screws, and machined slotted screw hardware.

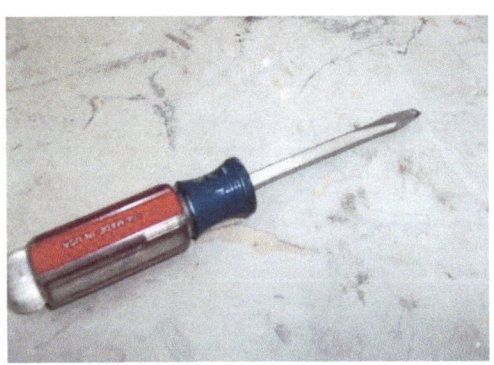

Phillips screwdriver

A Phillips or crossed-faced screwdriver is used to install, and remove "crossed" slotted wood screws and machined cross-slotted screw hardware.

Adjustable Mechanic wrenches 13"-10"-8"

Adjustable mechanic wrenches in various sizes can assist with removing nuts & bolts, tightening and loosening packing bonnets, and a variety of other uses.

Six-foot A frame ladder

A sturdy "A-frame" ladder made of fiberglass can assist with overhead space access, and also provide a sturdy work surface.

16'-0" Aluminum extension ladder

A sturdy aluminum extension ladder can assist with gaining access onto a building rooftop when necessary.

3 lb. Demo hammer

A 3-pound hand hammer is an excellent tool for dislodging, and loosening rusted equipment and material.

Sturdy work shoes (work boots preferred)
A sturdy leather work boots is an excellent choice for foot protection when working with equipment and tools.

Drain-plunger
An excellent tool for clearing small isolated clogs in drain p-traps and floor sink drains.

Comfortable work attire
Denim work pant, cotton long sleeve work shirt with your company name and logo above front left chest pocket (name and logo preferred not required) comfortable work cap.

Cell phone
Making and returning phone calls when needed. Being available when your specialized service is needed.

Remember you're selling professionalism.
"You can never over due professional"

- **Equipment**

Use of power equipment can be an excellent advantage in completing projects fast and efficient. Don't go cheap on quality here. Do not let the cost of tools and equipment frighten you and become a barrier between you and financial independence.
An average service charge is $85.00-$100.00 per hour starting. You will receive your investment back over and above your initial cost.

Do not empty your savings account buying a big-ticket equipment item. Purchase that special tool or equipment only when you need it to complete a work agreement that will require its use. Always include your material and equipment cost when submitting an estimate in order to complete a service arrangement without a price conflict. Also, include the cost of any special tools if needed when you prepare an itemized estimate.

Many plumbing tools and equipment items can be leased or rented at affordable rates from most industrial equipment rental suppliers and home improvement centers. A large equipment inventory is not necessary prior to starting your business.

Here is some power equipment I would recommend when starting your home repair drain-cleaning service:

Cordless 3/8" power drill
Used for drilling, boring, and anchoring applications

Reciprocating saw-all
Used for cutting wood studs, pipe, and various other construction related applications.

7/8" Circular saw
Use for various construction applications with wood, board, pipe, and other like material.

5/8" or 1/2" Power driven cable drain auger (DRAIN SNAKE)
Used to clean building drain pipe and sewer lateral waste pipe
form sizes 2"- 4" on residential and commercial application.

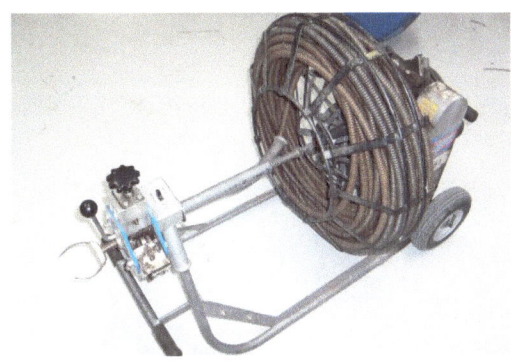

3/8" "Top-snake" Power driven drain auger Used to clear
clogged drains in waste branches sized from 1.5"- 2" on building
drainpipe. Lavatory fixtures, tub, shower, kitchen sink, lavatories, floor
drain and rain gutter drainpipes are easily accessible with the hand held
drain auger.

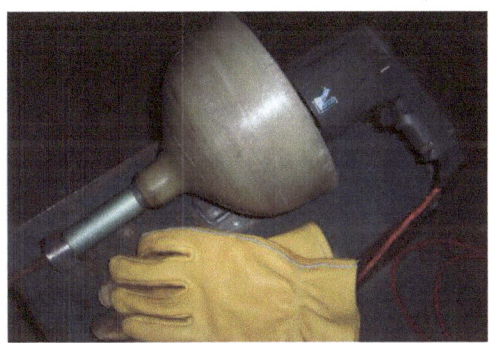

- **Closet Snake:**
A closet snake is used to access, and clean the internal trap of
water closet (toilet) and urinal fixtures.

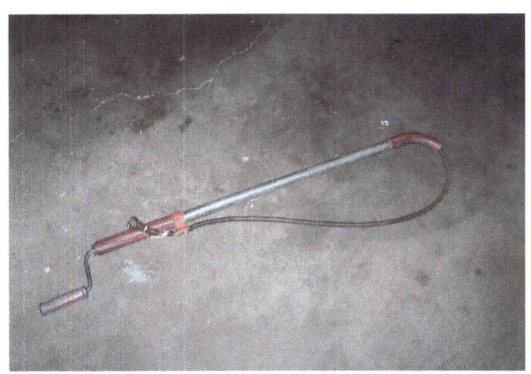

50'-0 extension cord
Used to connect into a remote outlet.

Materials

A specific code or schedule classifies some plumbing and building materials. They may have a factory applied identity color or classification reference for a specific application. Water pipe and sewer pipes have different recognized standards and application.

- Water pipe and connecting fittings used for potable water consumption should be manufactured with copper, brass, galvanize iron, poly vinyl chloride (PVC), chlorinated poly vinyl chloride (CPVC), bitimized fiber material, or other approved materials.

- Copper pipe can be identified by a factory applied color code that runs continuously on the developed length of the pipe exterior. Color identification is referenced as follows: Type K, green: Type L, blue: Type M, red: Type DWV (Drain waste & vent), yellow.

- All soft iron copper fittings used for potable water consumption should be copper, brass or galvanized iron.

- Pipes in length of 2" through 12" are referred to as "nipple" fittings. Nipple fittings are sized from: close, 2", 2 1/2", 3", 4", 5"etc. Size, material and lengths identify nipples.

- Building sewer pipe is made from a variety of materials. Cast Iron, ABS, Clay, PVC and Copper.

- The building sewer begins approximately (3) feet from the nearest fixture in front of a residential building, and should have a slope of not less than 2% or 1/4" grade per foot allowing sewage waste and gray waste to drain by gravity into the building sewer.

- Materials used for connecting plumbing service pipe are referred to as "fittings." Fittings can create branch offsets, stacks, slopes, clean out terminals, and other adjoining plumbing applications.

Familiarize yourself with the proper names and uses for each of the fittings I have detailed. Also, other fitting connectors like, unions, bushings, and valves will become common terminology you will soon be using.

I have identified several common fittings and their use:

- **Tee fitting:**
Used to make vertical and horizontal branch changes in a main pipe run.

- **1/4 Quarter-bend 90-degree elbow fitting:**
Can be sized with long sweep pipe and short sweep pipe elbows and is used to make vertical and horizontal offset in piping systems.

- **1/8 Eighth-bend 45-degree elbow fitting:**

Used to make a branch changes of forty-five degrees, ninety-degree and one hundred thirty-five degree offsets when used in a piping system.

- **Union connector fitting:**

Used to attach separated pipe branches or dissimilar metal connections where electrolysis may occur.

- **Coupling fitting:**

Used to extend pipe branches or extend pipes total developed length.

- **Pipe "Bell Reducer" fitting:**
  Used to increase or to reduce the size of a pipe branch.

- **Female iron pipe adapter (F.I.P) fittings:**
  Have an internal pipe thread that receives threaded pipe.

- **Male iron pipe adapter (M.I.P) fittings:**
  Has an external pipe thread that may be extended by use of a coupling fitting.

- **P-trap connector:**
  Used to provide water seal that will prevent sewer gases from entering the building premises.

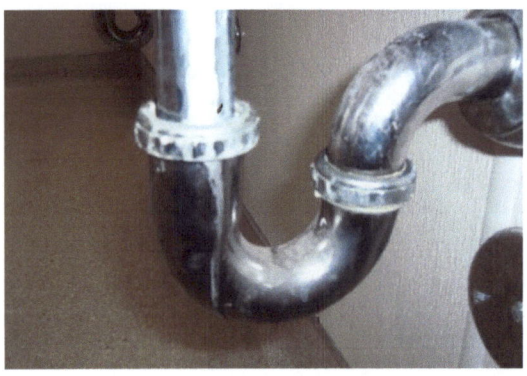

- **Combination wye and 1/8 bend or "comby" fitting:**
  Used in waste pipe branch and in horizontal drainpipe drain installation.

- **Wye fitting:**
  Used to make directional pipe changes or to connect a pipeline from an adjoining pipe branch.

- **Sanitary "Test-Tee" fitting:**
Used vertically with waste and soil pipe systems to provide an accessible clean out to the building sewer.

- **Sanitary tee fitting:**
Used to receive waste from the fixture trap arm and carries it into the building waste connection and out into building drain lateral.

- **Trap arm:**
The physical section of a fixture drain between the P-trap and the waste and vent connection.

- **Trap seal:**
The vertical distance between the crown weir of a p-trap and the top dip of a trap creating an air tight seal to avoid sewer gases from entering building space.

- **Crown Weir (Trap Weir):**
The lowest point in the horizontal waterway at the outlet of the p-trap.

- **Dip (of p-trap):**
The highest point of the cross sectional area of a p-trap. The bottom dip is lowest point in the internal cross section of the trap.

We will cover a variety of definitions and terms that are used in the plumbing code and trade reference in the second edition of "How to become a Journeyman Plumber" by Broad Vision Publishing, complete with sample test questions and answers.

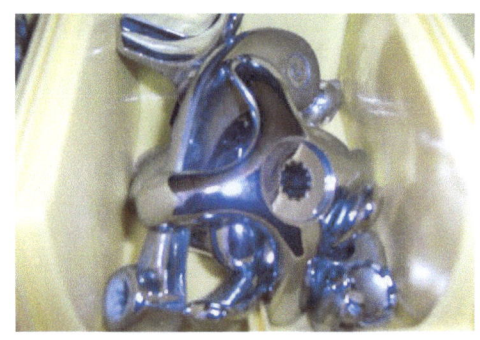

Basic Home Plumbing Maintenance

Inside the "Basic home plumbing maintenance" section, I will explain a number of techniques you can use to complete some home maintenance repairs and installations. You will learn how to identify and service clogged building drain problem.
How to install kitchen and bathroom faucets. Also, how to trouble-shoot other problems like leaking faucets, valves and hoses bibbs.

Before you can become a successful home repair business, it will first require that you learn how to properly identify repairs. The skill in making visual assertion of a repair is very important.

Review all of the information that is provided for you about the fixture you have purchased. This information will assist you in making the proper evaluation on the installation and maintenance of a particular fixture or appliance, if available. Prepare a detailed list of everything you will need to complete the job. Include all tools, materials, and labor cost that will be required to complete the project.

Never submit an estimate on any project without first preparing a detailed material list, including an approximate labor cost. Always attempt to complete your bid at cost! And on time!

Completing a project at cost can build positive customer associations, and assist in building a strong clientele network. "Word of mouth" referrals are major benefactors when you are establishing a home repair service. By delivering on time, and at cost, you will build a professional and honorable business reputation.

- **ESTIMATING**

When you prepare a project proposal, <u>**do not**</u> make a conscience decision to submit the lowest estimate. Trust in your accounting accuracy and project detail when you prepare a bid.

A poorly prepared, non-itemized estimate can be catastrophic when establishing a home repair business. The hardship of not maximizing your time by properly preparing an itemized estimate can create huge deficient profit margins, and financial instability.

In most cases, the lowest estimate does not insure that you will get work anyway. In some cases, you will be better off not getting some jobs, than you would be if you were to loose money on labor, material and profit. Remember **"Time Is Money"** now!

The first rule of business is you are not going to make every sale. To your advantage, there is an enormous shortage of qualified home repair service technicians, drain-cleaning technicians (apprentice plumber), and a lot of work in the related trades right now.

There will be times when the work petitions can become demanding, and a complicated balance. When you are in a business that market a technical trade knowledge and skilled competence you will find lots of work to be done. Prepare yourself. Always plan ahead look forward to the next project on the job schedule.

The biggest mistake a lot of new businesses make in the conception stage of launching a home repair service is, they attempt to get rich with their first service repair job. Good service will generate BIG PROFITS! Consistency will build a strong financial future.

While constructing your new home repair and drain-cleaning service you should include a business blueprint program. This program allows you to reinvest funds back into your company. By investing a small percentage of the gross income back into your home repair service you can accumulate a lot of new prospective accounts. It can also assist you with purchasing advertisement ads, tools, equipment, and enrolling into some professional trade classes.

In the beginning you may experience some feelings of anxiety when you receive inquires requesting your home repair service. You may perceive that too many things are taking place all at once. That is a common reaction when embarking upon a new profession. From medical professionals to a shoe sales person, you may become overwhelmed by your newly acquired position. As a business owner the associated responsibilities can create some emotional apprehension. You can question yourself as to whether or not you made the right decision? "Can I complete the job I've promised the customer with skill and competence?" Etc. etc. etc. Well, my answer to you is "Yes you can and yes you will."

The success of your business venture will ultimately be the ability to complete a project. It will not be how fast you can do it. Approach your work without an emotional connection affixed to it. Do not get excited like some homeowners may become. You can professionally assess what will be necessary to make competent skilled repairs and earn substantial profits by doing it.

Secondly, you will need to know how to respond when dealing with a panicked customer. **Never over react**! You are there because **you're** the skilled professional and the "problem solver" that is why you can earn $85.00-$100.00 per hour confidently! Always, maintain a high level of professionalism.

By nature, we can often lead our minds down some pretty strange and difficult paths when we're attempting to resolve problems.

Anxiety can convince us that we are just not mechanically capable to complete the challenges of making repairs correctly.

Because I am a professional plumber, I have observed some really interesting variations on what *not* to do. Lots of people just freak out once they are "outside of their comfort zone." Although my research is not based on any particular scientific study, it just appears to be that way.

Once a proper evaluation is made on any repair the job can be completed with a proficient level of confidence and ease. By following a few simple procedures, I will show you exactly how to do things properly when resolving a plumbing or drain-cleaning dilemma.

Now that you have prepared to begin your journey toward financial independence by acquiring a few tools, soliciting perspective customers, and reading all of the information you will need to get yourself started working. I will show you how to get started in your new and exciting professional career with skill and confidence.

How to service clogged drains:

Let's simplify the cause of most clogged drains, and equipment used to remove the hysteria of having standing water inside a kitchen sink, or lavatory fixture. Here are some things that are not good for plumbing drains and p-traps: cooking grease, oils, fruit and vegetable peels, coffee grounds, hair, soap deposits, shaving cream and other materials may paste or solidify inside a drain pipe.

Using a standard drain plunger to remove a drain obstruction can easily resolve many backed-up drainage problems. A drain plunger uses force to push the clogged blockage through the fixture p-trap and into the building drain. Although some stoppages may require the use of power auger equipment to satisfy a repair, it is always advisable to begin with the simplest approach first to complete a repair.

If a power drain auger (drain snake) is required to complete a clogged drain problem start by observing all safety warnings and operation instructions. Your local equipment rental supplier can often show an individual with no prior experience, proper use of drain cleaning auger equipment. Most drain obstructions requiring the use of a power drain auger will often be caused by tree roots, heavy grease, oil products, children toys, and other foreign properties that may enter a waste branch or the building drainage system.

When preparing to repair a building drainpipe system, thoroughly observe the surrounding areas. Also, look for all clean-out (C.O.) access locations. Try to locate the nearest accessible location to pick up electrical power, and a convenient place to position drain equipment.

Buildings and structures including covered or uncovered, porches, steps, breezeways, patios, carports, walkways, and similar structures should have an accessible clean out within 2' -0" outside of the building. Its closest fixture exiting the building drain should connect into the building sewer connection. The clean out sweeps should be installed outside the building. The clean out should be located at the lower end of the building drain and extended to finished grade.

(Residential) Main-Line Drain Cleaning:

Residential main-line drains can be cleaned by cabling through the building clean out, or the main fixture vent. Clean out access should be installed to allow cleaning in the direction of flow with soil or waste pipe. Clean-outs installed under concrete or asphalt should have an accessible yard box that should be inline with the building drain, or building sewer it serves. Clean out plugs should have a raised square head or an approved counter sunk rectangular slot for removal.

Once you have assessed the repair and determined what will be necessary to complete the job i.e. tools, equipment, and labor. The use of a ¾" power drain auger equipped with an appropriate cutter will assist in removing difficult drain obstructions when clearing most mainline stoppages.

1. Remove clean out plug, (You may notice liquid or waste spilling over top of C.O. access when first opened.)

2. Extend approximately 2'-0" to 3'-0" of auger cable then place it inside clean out access before engaging power to auger equipment.

3. While wearing the drain cable "ugly gloves" engage foot pedal as you begin feeding cable into waste pipe at moderate intervals. You will notice the water level may rise and fall slightly as cable is fed through the pipe. Once you clear the location causing drain stoppage, standing water will rapidly be released and discharge through sewer lateral. The waste branch can require as much as 75'-0" or more auger cable before it reaches the problem inside a "Main-Line" clogged waste branch. Resistance to the auger cable while feeding it through a blocked drainpipe is a good indicator of the problem location. Do not rush the process, once the location is reached that is causing the blockage the drain branch should clear.

4. After repairs are completed, gather all tools and equipment. Also, spot clean your work area before presenting your customer with an invoice for payment. Always promote yourself and business professionally.

The minimum service charge for a "Main Line" drain cleaning in my geographic area to date, is $165.00 for the first hour plus $85.00 for every hour thereafter. So, at 1.75 hours you may earn $250.00 in less than 2 hours of work. Now, you can see just how lucrative your time has just become!

All estimates are given for the purpose of an example only, and are not intended to be quotes or bids to assist in any manner, nor are they intended to solicit or procure work, or any contract agreement.

Water closet stoppage (Toilet):

The water closets and urinal have an integral p-trap design inside of the plumbing fixture. Water closets are mounted to the building floor with brass retaining bolts for base connections, or with carrier fittings for wall mounted fixtures hung by and a wax seal gasket or felt material. Also, urinals are attached using wall hangers and wax or felt seal gasket. When servicing either water closets or urinal fixtures use a "Closet snake." It will more likely produce a favorable solution to stubborn clogs.

The closet snake has a curved shaft at its base that maneuvers around bends allowing easier access through fixtures p-trap and connected waste branch to clear or remove drain obstructions. You may also attempt to resolve toilet and urinal clogs with a drain plunger.

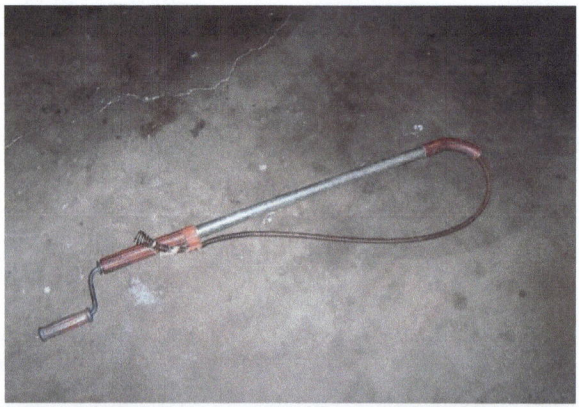

Kitchen sink stoppage:

Kitchen sink clean outs are usually within close proximity of the fixture or near the fixture drain branch. If necessary, you may cable clean the connected waste branch through an open clean out access, or through fixture trap arm, and use a portable hand held power auger or " Top snake" as referred to in the plumbing trade. Top snakes are equipped with a small diameter auger cable that will allow accessing 1 ¼" to 2" drain branches smooth and easy to reach. You may also try using a standard drain plunger.

Lavatory sink stoppage:

Lavatory sink clean outs are usually within close proximity of the fixture, or near the fixture drain branch. If necessary, you may cable clean the connected waste branch through an open clean out access, or through fixture trap arm, and use a portable hand held power auger or " Top Snake" as referred to in the plumbing trade. Top snakes are equipped with a smaller diameter auger cable that will allow accessing 1 ¼" to 2" drain branches smoother and easier to reach obstructions. You may also try using a standard drain plunger.

If it is necessary to cable clean a kitchen sink or lavatory sink to clear a blockage:

1. Access and remove the clean out plug.

2. Insert approximately 2'-0" to 3'-0" of auger cable into the test-tee fitting, or connected drain branch.

3. Plug in drain machine equipment into power source and slowly hand feed drain cable into the isolated drainpipe.

Most isolated drain clogs are between the fixture drain and building drain that may not warrant an extensive cable-cleaning project to resolve drainage problem.

If any concealed trap, drainpipe, or soil pipe become defective while making a repair, it may become necessary to remove and replace all defective parts with new materials. In some circumstances a permit may be required if a fixture or an appliance is replaced.

Since you are not a licensed journeyman plumber or plumbing contractor, any required permits will have to be obtained by the homeowner or their responsible representative. First, discuss any additional cost that may apply to any work being performed with the customer or responsible party for payment. Second, never perform any additional work without authorization. Third, list any addition to material supplies and labor cost that may apply with your billing receipt.

Bath Tub and Shower Drains:
Bath tub and shower drains are two common and profitable service repairs. Drain-cleaning companies make routine repairs to tub and shower drains often. Because of the daily use of both fixtures they can start to drain slow and sluggishly. Our bodies shed hair, dead skin, body oils, dirt and soap debris. Over time those contaminates can cause plumbing drain fixtures to drain irregularly, that may require a need to cable clean the fixtures p-trap and waste branches.

To service a bathtub drain:

1. Remove the waste overflow cover or the "Trip-Lever" cover located at top of the drain inlet. Look for a chrome cover plate with two slotted retaining screws holding it in place (slotted screws usually).

2. Using your hand held "Top-Snake" power auger, insert approximatly12"-16" of cable into the overflow pipe. Plug in power drain equipment and engage power.

3. Hand feed power auger cable through bathtub overflow pipe, and into the fixture drain waste branch at a moderate interval. Run out roughly 15'-0" to reach the 3" building drain.

4. Extract cable and reassemble overflow cover. Test as needed to assure blockage has been resolved. Once drain concerns have been repaired, spot clean work area, and submit invoice for payment. Include all material and labor where applicable.

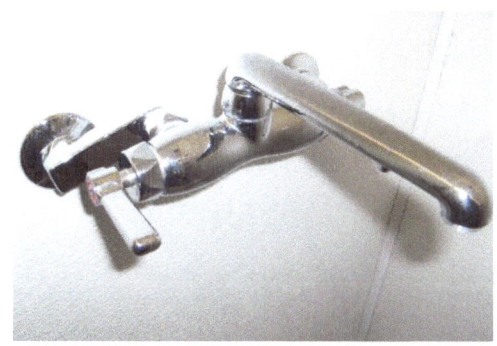

How to Install a Kitchen Faucet:

Installing a kitchen faucet is not as complicated as one might think. With a vast variety of faucet design and finishes to choose from, it will probably become more of a challenge assisting the customer in finding the perfect faucet. You can find a lot of great bargains on quality faucet fixtures from your local home improvement center, and most plumbing part suppliers. The price range on faucets can vary quite considerably from very inexpensive; to cost that can amount to a small mortgage loan. Assist your customer with good quality and selection when choosing.

Weather it is a wall mount faucet or a counter mount fixture; the same preparation will apply prior to installation.

1. "Shut-off" water to faucet first by locating the hot and cold angle stop shut off valve (S.O.) they should be located beneath the cabinet space directly under the kitchen sink. If there are no accessible (S.O.) valves you will have to isolate the water supply at the main entrance of the building, or at the main water valve entering the dwelling. There should be an accessible "Gate Valve" or "Ball Valve" to close off potable water supply.

2. After the potable water is shut off you can begin to disassemble the old faucet. You should find two slip nut retainers at the base of the hot and cold sides of the faucet. The retainer nuts are used to mount the faucet to the wall and to connect into the water distribution pipes. By loosing the two retainer nuts the faucet should become dislodged allowing the faucet to easily be removed.

After the old fixture has been successfully removed, clean up any rust or corrosion debris that may be left behind from the old discarded faucet. Look at both hot and cold supply nipples that extend through the wall to ensure they are free of any rust or corrosion. Now, if there is any sign of visible corrosion around either of the supply nipples, it is recommended that you replace one or both water service nipples. You can prevent most problems that may occur. Such as leaks, dripping connections and other nuisances that can impede on a successful installation job. Assure all connections and fittings are in good order.

Now that you have a clean watertight workspace wrap both hot and cold supply nipples with a minimum 1/2" Teflon pipe tape to prevent any leaks or seepage that may occur when the building water is turned back on. Apply pipe tape around the supply nipples extending through the flange connections on both hot and cold nipple connections. Follow all of the owner manual information provided with the new faucet to ensure that proper installation practices are followed. After the installation is completed, test the operation of the new apparatus. Check for leaks or other malfunctions. If any modifications are required make corrections and cleanup work area and tools prior to submitting an invoice for payment. Also, save any information that may be applied to fixture warranty.

How to Install a Counter Top Kitchen faucet:

Prior to beginning installation observe the existing fixture and the counter top work surface. Look around the area and identify all mechanical connections that can be disconnected either in stages or as a unit. Awareness of what will be needed before you begin to disassemble the old faucet will assist you in a prompt and complete job.

"Shut-off" water to faucet, by locating the hot and cold angle stop shut off valve (S.O.) they should be located beneath the cabinet space directly under the kitchen sink. If there are no accessible (S.O.) valves you will have to isolate the water supply at the main entrance of the building, or at the main water valve entering the dwelling. There should be an accessible "Gate Valve" or "Ball Valve" to close off potable water supply.

 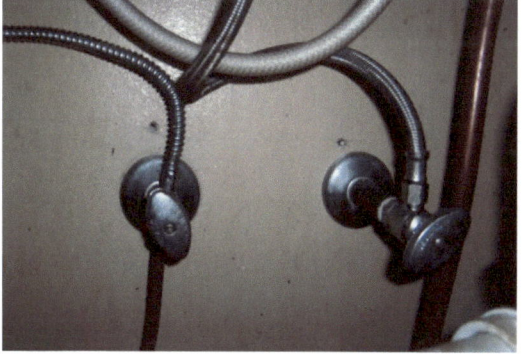

After the potable water is shut off you can begin to disassemble the old faucet. You should find two slip nut retainers at the bottom of the hot and cold sides of the faucet. The retainer nuts are used to mount the faucet to the counter top and to connect into the water distribution pipes. First, by removing the two water supply connectors, you can access the retainer nuts located at the bottom of the faucet allowing you to easily dislodge the faucet and remove it.

After the old fixture has been successfully removed, clean up any rust or corrosion debris that may be left behind from the old discarded faucet. Look at both hot and cold supply nipples that extend through counter top deck to ensure they are free of any rust or corrosion. Now, If there is any sign of visible corrosion around either of the supply nipples, it is recommended that you replace one or both water service nipples. Also, you may consider replacing both hot and cold water supply connectors. Check sizes of both connectors before you purchase materials. Most common size for faucet to supply connections are ½"x ½" or ½" x 3/8" so check to be sure.

You can prevent most problems that may occur like leaks, dripping connections and other nuisances that can impede on a successful installation job by assuring all connections and fittings are in good order.

Apply "Plumber's Putty" around the hot and cold nipples at the base of the new faucet before placing it through the sink hole openings and securing it in place. Plumber's putty will prevent water and liquids from seeping through small gaps around the base of fixture. Follow all of the owner manual information provided with the new faucet to ensure that proper installation practices are followed. After the installation is completed, test the operation of the new faucet. Check for leaks or other malfunctions. If any modifications are required make corrections and cleanup work area and tools prior to submitting an invoice for payment. Also, save any information that may be applied to fixture warranty.

How to Install a Lavatory Faucet:

Lavatory faucets can have eccentric detail and operation components designed into the fixture. The control and function of any faucet is to provide tempered water on demand, close with a positive seal when faucet is shut off prevent any leaks or drips, and should be free of any lead base or compound. "Lead free" is defined as having no hazardous lead material used in the fabrication of the apparatus. When installing lavatory faucets follow the same process applied with a counter top faucet application. Most lavatory faucets are prefabricated with a 4" rough-in measurement, which means the space between each supply nipple is four inches apart. However, with a vast variety designer bathroom fixtures and components available now, you can purchase a specialty ordered faucet with an 8" rough-in. Always measure the sinks top opening and confirm with the customer when making a specialized order, some venders may not allow returns.

"Shut-off" water to lavatory faucet, by locating the hot and cold angle stop shut off valve (S.O.) they should be located beneath the cabinet space directly under the lavatory sink.

 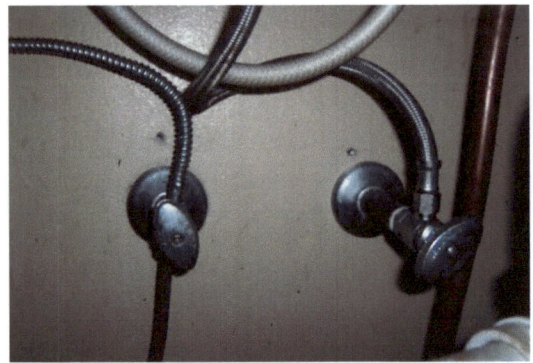

After the potable water is shut off you can begin to disassemble the old faucet. You should find two slip nut retainers at the bottom of the hot and cold sides of the faucet. The retainer nuts are used to mount the faucet to the counter top and to connect into the water distribution pipes. By removing the two water supply connectors, you can access the retainer nuts located at the bottom of the faucet allowing you to easily dislodge the faucet and remove it.

After the old fixture has been successfully removed, clean up any rust or corrosion debris that may be left behind from the old discarded faucet. Look at both hot and cold supply nipples that extend through the wall to ensure they are free of any rust or corrosion. If there is any sign of visible corrosion around either of the supply nipples, it is recommended that you replace one or both water service nipples. Also, you may consider replacing both hot and cold water supply connectors. Check sizes of both connectors before you purchase materials. Most common size for faucet to supply connections are ½ "x ½" or ½" x 3/8" so check to be sure.

Apply "Plumber's Putty" around the hot and cold nipples at the base of the new faucet before placing it through the sink hole openings and securing it in place. Plumber's putty will prevent water and liquids from seeping through small gaps around the base of fixture. Follow all of the owner manual information provided with the new faucet to ensure that proper installation practices are followed. After the installation is completed, test the operation of the new faucet. Check for leaks or other malfunctions. If any modifications are required make corrections and cleanup work area and tools prior to submitting an invoice for payment. Also, save any information that may be applied to fixture warranty.

How to Repair Leaking hose Bibbs:
Hose bibbs can often become worn over time from constant use and high demand. Leaking or broken hose bibb repairs can be as simple as replacing a defective seat washer to resolve most problems with a bibb faucet, or if necessary, you can replace the

entire fixture with a relatively inexpensive cost to you. A $3.99 hose bibb can earn you a substantial profit margin when you consider the cost per hour you are earning, for example: $85.00 per hour, plus $3.99 in material cost can net $88.99 for a twenty-minute service repair. Profits are even greater when you keep in mind the cost of an average hose bibb washer is far less than the cost of a replacement part.

1. "Shut-off" water to the hose bibb, by locating the building shut off valve (for residential) dwellings. The building water shut off valve should be located outside of the structure where the water service utility pipe enters the building or at the water meter utility box.

2. After water is shut off, remove the defective part and observe the condition of the fixture. If the hose bibb appear to be free of any damage or corrosion, simply replace the integral rubber washer inside.

3. Remove the handle. Take out the set-screw on the top of the handle.

4. Loosen the retaining bonnet nut that secures the valve stem to the hose bibb body. Turn valve stem in a counter-clock wise direction to remove stem from the hose bibb body.

5. Replace the washer at the bottom of the hose bibb with a new washer and check to insure valve seat is not pitted. If there is noticeable pitting around the valve seat it should be replaced. Also, apply a small wrap of teflon packing tape beneath the

bonnet retainer nut when reassembling the unit to prevent any seepage that may occur around the top of fixture handle.

6. Reconnect hose bibb to its service supply pipe and turn on the building water from where it was isolated. Check for any leaks or malfunction prior to submitting your invoice for payment.

The information I have provided in this manual can benefit individuals with an interest in pursuing a career in the plumbing drain cleaning profession, plumbing maintenance and operations or in a building trade occupation. I have attempted to outline some of the most basic trade functions that may assist you with an immediate start up of a small business, while continuing to develop your career goals and objectives.

With a detailed plan you can be on your way toward financial independence and entrepreneur ownership in a very short time. Plumbing have truly been a rewarding vocational decision for me over the years and I am excited to share my knowledge and the information you can use to get started on your way TODAY!

Good luck in all of your professional trade pursuits and accomplishments.

Look forward to the next addition from BROAD Vision Publishing: How to become a "Certified Journeyman Plumber," and preparing for your "C-36 Plumbing Contractor Exam" Complete with example test questions and answers. Coming soon!

Sincerely,
BROAD Vision Publishing

Congratulations! And welcome to BROAD Vision Publishing,

You have made a great decision that can benefit you for a lifetime, by joining countless other men and women who are now building careers as non-licensed plumbing drain cleaning technician, installation and repair businesses. You will soon be discovering one of the most lucrative and personally rewarding career choices you can ever make.

Weather you are planning on becoming a drain cleaning service technician, licensed journeyman plumber, C-36 state licensed plumbing contractor, or just a pipe and supply vendor, this billion dollar plumbing industry can help you build profitable resources and a rewarding career.

Commercial and Residential Developers are constantly pursuing qualified union plumbers. State and City government, county agencies, schools, hospitals and homeowners alike will at some point be in need of the services that a professional drain cleaning technician, installation and repair business can provide. The earning potential for a licensed plumber is virtually endless. Some licensed plumbers can earn $65,000-90,000 annually.

Major drain cleaning businesses can earn over $150,000 or more annually. I will show you how to earn $100.00 an hour as a non-licensed commercial and residential drain cleaning technician, installation and repair business.

Start now! For $29.95 and discover for yourself endless financial opportunities available to you right now!